INTERIOR DESIGN GALLERY

矩阵纵横设计作品精选

M matrix

主编：矩阵纵横设计团队   以矩为方圆·万物皆为阵

江苏人民出版社

MATRIX Profile

Founded in 2010, MATRIX is a design brand aiming to provide high-end design services. The leading designers of the matrix team have all gained longtime working experience of providing design services to the domestic well-known property developers, and have built a long-term stable and trustable cooperative partner relationship with our clients. We are looking forward to long-term growing and developing together with the old and new clients that trust us. We would like to express our gratitude hereby!

矩阵纵横为2010年成立的旨在提供高端设计服务的设计品牌，团队主创人员皆有多年服务于国内知名房产开发商的工作经验，并与客户建立了长期稳定、相互信任、共同发展的合作伙伴关系。对于信任我们的新老客户，我们愿意与其共同成长和发展。在此我们深表谢意！

矩阵简介

Matrix Interior Design
Show Flat And Sales Center
Club House Restaurant Spaces
Commercial
Office Spaces
Hotels

www.matrixdesign.cn © Copy Right

CATALOG

Show Flat 101, Purple Garden House, Anhui Huadi Real Estate 安徽华地紫园-洋房101户型	006/007
Show Flat 501, Purple Garden House, Anhui Huadi Real Estate 安徽华地紫园-洋房501户型	024/025
Show Flat 1101, Purple Garden House, Anhui Huadi Real Estate 安徽华地紫园-洋房1101户型	038/039
GARA Boutique, Shenzhen GARA世界名牌精品店（深圳）	060/061
Sales Center of Victory Nan Cheng, Grace From Heaven Real Estate, Chengdu 天合地产（成都）凯旋南城售楼中心	068/069
Show Flat, Wongtee Xi Yuan Villa, Dameisha, Shenzhen 深圳大梅沙皇庭熙园别墅示范单位	076/077
No. 7 Villa, Tian Yue Wan, Shenzhen Gemdale Real Estate 金地深圳天悦湾7号楼别墅空间	090/091
No.3 Villa, Tian Yue Wan, Shenzhen Gemdale Real Estate 金地深圳天悦湾3号楼别墅空间	106/107
Sales Center of Wonder Town, Vanke Real Estate, Chongqing 万科地产（重庆）缇香郡销售中心	118/119
Show Flat D2, Joy Town, Wuxi 无锡复地悦城D2户型	132/133
Peng Cheng Office, Shenzhen 彭成实业有限公司办公室（深圳）	144/145
APYRITE 碧玺	154/155

目录

Left Bank of Bordeaux 波尔多左岸	166/167
Hair Q Salon, HK 香港Hair Q Salon 形象中心	182/183
Switzerland Royal Anti-Ageing Company Reception Club, Shenzhen SRN瑞士皇家抗衰老公司(深圳接待会所)	186/187
Show Flat, Uptown, Chongqing 重庆复地上城示范单位	192/193
Joy Town, Shenzhen 深圳中央悦城别墅	200/201
Penthouse, St-Maurice Garden, Shenzhen 圣莫丽斯顶层行宫	212/213
J Club, Verakin New Park City 同景地产国际城J会所	228/229
Office of Vanke Real Estate, Chongqing 万科地产（重庆）办公楼	232/233
Vanke Health Club House 万科地产（重庆）天琴湾次会所	246/247
Tian Qin Wan Golf Club, Vanke Real Estate, Chongqing 万科地产（重庆）天琴湾高尔夫会所	256/257
FRAME Special Interview FRAME杂志专访	292/293
Process 心路历程	294/295

Show Flat 101, Purple Garden House, Anhui Huadi Real Estate

安徽华地紫园-洋房101户型

Project Contents: Show Flat
Design Area: 375 m²
Hardware Decoration Cost: RMB 3000/m²
Soft Decoration Cost: RMB 3000/m²
Main Materials: Stone, Carpet, Wallpaper, Gilding Cloth, Leather Soft Package, Rose-gold Stainless Steel, White Grain Paint, Mosaic, etc.
Time of Design: Jan, 2011
Completion Time: Oct, 2011
Design Specification: The foreign-style house is a compound structure consisted of first floor and underground basement, Red wine is the theme for this model house, the water which acts like a waterfall and runs through the first floor and basement has been led into the house, thus, the concept of elite housing is stressed by passing through the hall into the inner chamber, attractive scenes have been borrowed at different floors. A high quality space which can match the space of villa has been created at the bottom of the house.

项目性质：样板房
设计面积：375平方米
硬装造价：3000元/平方米
软装造价：3000元/平方米
主要材料：石材、地毯、墙纸、烫金布、皮革软包、玫瑰金不锈钢、白色显木纹漆、马赛克等
设计时间：2011年01月
竣工时间：2011年10月
设计说明：鉴于洋房的一楼及带有下沉庭院的负一楼是复式结构，设计师以红酒为主题演绎全新的"亲地别墅"样板房。入户时即开始通过"泗水归堂"的概念将整个小区园林的水系引入本户型，通过流水瀑布的手法贯穿两层空间，并在空间关系上强调了"登堂入室"的豪宅观念，同时层层借景，步步引人入胜。在一个原本只是洋房的底层空间打造出一个堪比别墅的高品质空间。

MATRIX DESIGN

MATRIX DESIGN

First Floor Plan 一层平面布置图
01.客厅
02.西厨、酒吧
03.餐厅
04.内玄关
05.主卧室
06.次主卧室
07.卧室
08.入户前院
09.门厅
10.工人房
11.前庭院
12.休闲庭院

B1 Floor Plan 负一层平面布置图

01.红酒雪茄游戏区
02.台球娱乐区
03.视听室
04.楼梯造景
05.下沉流水造景
06.储藏室
07.电梯厅

MATRIX DESIGN

Show Flat 501, Purple Garden House, Anhui Huadi Real Estate

安徽华地紫园-洋房501户型

Project Contents: Show Flat
Design Area: 150 m²
Hardware Decoration Cost: RMB 3000/m²
Soft Decoration Cost: RMB 3000/m²
Main Materials: Oak Veneer, Ebony Veneer, White Travertine, Wood Flooring, Textile Wallpaper, Artificial Leather, Black Mirror, Black Steel, etc.
Time of Design: Jan, 2011
Completion Time: Oct, 2011

项目性质：样板房
设计面积：150平方米
硬装造价：3000元/平方米
软装造价：3000元/平方米
主要材料：橡木饰面、黑檀木饰面、白洞石材、木地板、墙布、人造皮革、黑镜、黑钢等
设计时间：2011年01月
竣工时间：2011年10月

Design Specification:

The owner requests a combination of display and functional space for houses which consist of the most parts of the project, thus, impression and communication of various spaces have been paid a special attention during design, sunlight has been introduced into the interior, while beige, grey and black are adopted as the main colors to avoid a chaotic sense of colors and a smell of fashion has been added to the natural and ecological life space.

设计说明:
鉴于该户型是洋房数量最多的户型,业主要求空间展示与实际使用功能相结合,设计师在设计时延续了建筑的空间观念,着重处理空间的层次与互通关系,将光线引入室内,同时尽量避免出现过多繁复的造型,仅以米、灰、黑等色作为空间的主色调,为自然环保的生活空间中注入了一股时尚的气息。

MATRIX DESIGN

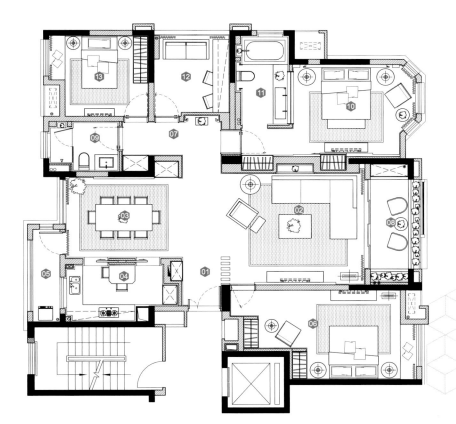

Ground Floor Plan 平面布置图
01.玄关区
02.客厅
03.餐厅
04.厨房
05.生活阳台
06.客卧
07.过廊
08.客厅阳台
09.公卫
10.主卧
11.主卫
12.书房
13.客卧

MATRIX DESIGN

MATRIX DESIGN

Show Flat 1101, Purple Garden House, Anhui Huadi Real Estate

安徽华地紫园-洋房1101户型

Project Contents: Show Flat
Design Area: 240 m²
Hardware Decoration Cost: RMB 3000/m²
Soft Decoration Cost: RMB 3000/m²
Main Materials: Red Travertine, Yellow Travertine, Oak Veneer, Wood Flooring, Wallpaper
Time of Design: Jan, 2011
Completion Time: Oct, 2011

项目性质：样板房
设计面积：240平方米
硬装造价：3000元/平方米
软装造价：3000元/平方米
主要材料：红洞石材、黄洞石材、橡木饰面、木地板、墙纸
设计时间：2011年01月
竣工时间：2011年10月

Design Specification:
What kind of space it could be for a compound foreign-style house on the top floor? A brand new mansion-like space has been taken on when a wide open view has been stressed and vertical circulations have been modified. A private space of master room has been developed and a storage space has been created while the requirement of original functional space has been met. The old shanghai style has been adopted, culture of mansion and full-bodied feeling of humanities have been vividly explained.

设计说明：
一个带有斜屋顶的复式顶层洋房空间会变成什么样子？在设计师的改造下突出了顶层空间的开阔视野，并且在调整立体交通动线后，一个以"顶层公馆"为主题的全新的空间展现在我们的眼前。在满足原有功能层次的前提下，又开发出了极具私密感的主卧空间，并且在巧妙利用斜屋顶后出现了第三层的储物空间，设计师采用老上海的风格进行了装饰演绎，突出了公馆的文化概念及浓浓的人文情怀。

MATRIX DESIGN

MATRIX DESIGN

MATRIX DESIGN

046/047

MATRIX DESIGN

First Floor Plan 一层平面布置图
01.客厅
02.餐厅
03.玄关
04.景观楼梯厅
05.厨房
06.主卧室
07.次卧室
08.景观阳台
09.客卫

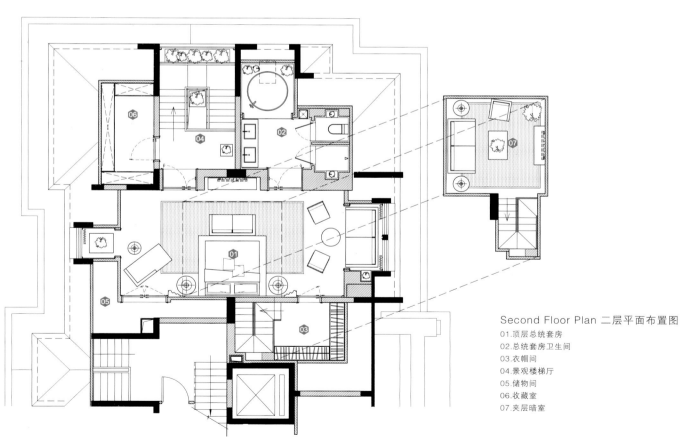

Second Floor Plan 二层平面布置图
01.顶层总统套房
02.总统套房卫生间
03.衣帽间
04.景观楼梯厅
05.储物间
06.收藏室
07.夹层暗室

MATRIX DESIGN

MATRIX DESIGN

GARA Boutique, Shenzhen

GARA世界名牌精品店（深圳）

（陈列宝个人作品）

Project Contents: Retail & Demonstration Area
Design Area: 400 m²
Hardware Decoration Cost: RMB 3000/m²
Soft Decoration Cost: RMB 3000/m²
Main Materials: Black Brick, GRG. Automotive Enamel, Mirror Finished Stainless Steel, etc.
Time of Design: Jun, 2010
Completion Time: Dec, 2010
Design Specification: Located in the beautiful Shenzhen Dameisha Outlets Mall, GARA is a fashion boutique including many world top brands such as GUCCI, PRADA, MIUMIU, DIOR, FEDI and etc.
Designer takes the flower as inspiration, and extends the idea to vase, petals, buds and other related elements for the design. The usage of pure white color and black background highlights the whiteness, creates spotless shopping environment with noble temperament and provides a harmonious and unified display space to all the unique fashionable boutique brands.

项目性质：零售及展示空间
设计面积：400平方米
硬装造价：3000元/平方米
软装造价：3000元/平方米
主要材料：黑色亮面砖、高强度玻璃纤维树脂、汽车漆、镜面不锈钢等
设计时间：2010年06月
竣工时间：2010年12月
设计说明：GARA世界名牌精品店位于美丽的深圳大梅沙奥特莱斯购物村，是一间集合GUCCI，PRADA，MIUMIU，DIOR，FEDI等世界顶尖品牌的时尚精品折扣店。
设计师以花为设计灵感，延伸出花瓶、花瓣、花蕾等设计元素，纯白色调搭配黑色的大背景，更加突显空间的洁白无瑕，营造出一尘不染的尊贵的购物环境，也给独具个性的各大时尚精品品牌提供了和谐统一的展示零售空间。

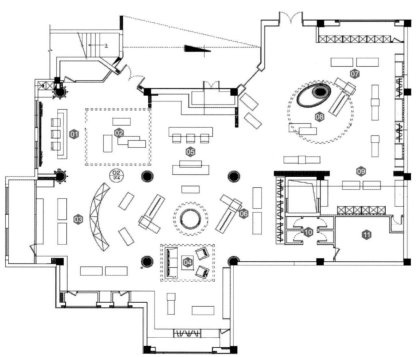

Ground Floor Plan 平面布置图

01.前台接待
02.接待休息区
03.女装专卖区
04.男装专卖区
05.眼镜专卖区
06.名包专卖区
07.名鞋专卖区
08.名鞋展示区
09.饰品区
10.试衣室
11.库房/办公室

MATRIX DESIGN

Sales Center of Victory Nan Cheng, Grace From Heaven Real Estate, Chengdu

天合地产（成都）凯旋南城售楼中心

Project Contents: Sales & Demonstration Area
Design Area: 1000 m²
Hardware Decoration Cost: RMB 2000/m²
Soft Decoration Cost: RMB 2000/m²
Main Materials: Black Jade, Castle Gray, Luo Manakin, BaLunke Gold, BaLunke Gray, Multi-layer Parquet, Black Mirror Steel, Black Ebony Veneer, Gray Soft Package, etc.
Time of Design: Mar, 2011
Completion Time: Sep, 2011

项目性质：楼盘销售展示
设计面积：1000平方米
硬装造价：2000元/平方米
软装造价：2000元/平方米
主要材料：黑海玉、古堡灰、罗曼金、巴伦克金、巴伦克灰、多层实木复合地板、黑镜钢、黑檀木饰面、灰色软包等。
设计时间：2011年03月
竣工时间：2011年09月

Design Specification:
This project shows calm and luxurious features of the art deco style. Groined ceiling, walls with simple-style stone modeling and the floor covered by gorgeous dazzling parquet match with unchained-style luxury chandelier made of copper, which creates a gorgeous and luxurious temperament and fully shows the art deco style.

设计说明:
该项目设计体现了art deco风格沉稳奢华的特点：井字形天花、硬朗简约的墙面石材造型和地面华丽炫目的拼花，配合造型奔放的铜制奢华吊灯，营造出大气简约、华丽奢华的空间气质，将其风格特征表现得淋漓尽致。

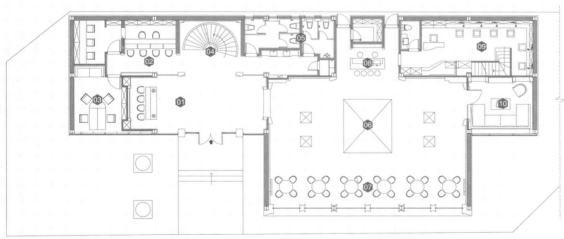

First Floor Plan 一层平面布置图
01.接待前台
02.缴费区
03.销售办公室
04.楼梯
05.卫生间
06.模型展区
07.洽谈区
08.水吧区
09.办公室样板房一层
10.VIP洽谈室

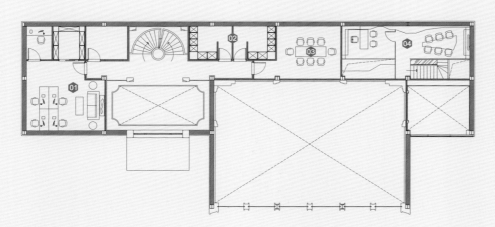

Second Floor Plan 二层平面布置图
1.物业办公室
2.更衣室
3.会议室
4.办公室样板房

Show Flat, Wongtee Xi Yuan Villa, Dameisha, Shenzhen

深圳大梅沙皇庭熙园别墅示范单位
(王冠个人作品)

Project Contents: Villa Show Flat
Design Area: 680 m²
Hardware Decoration Cost: RMB 3000/m²
Soft Decoration Cost: RMB 3000/m²
Main Materials: Blue Glass, Hard Coated Fabric, Ariston White Stone, White Brushing Lacquer, etc
Time of Design: Jun, 2009
Completion Time: Mar, 2010

项目性质：别墅样板房
设计面积：680平方米
硬装造价：3000元/平方米
软装造价：3000元/平方米
主要材料：蓝色玻璃、硬包面料、雅士白石材、白色手扫漆等
设计时间：2009年06月
竣工时间：2010年03月

Design Specification:
Considering the speciality of the project, the designer plans it in the key concept of Private Hotel and Heart of Ocean, which leads to the whole penetration of different bluish greens that the ocean presents in a day, with white color that stands for the sunshine and beach serving as the contrast. Thus, the project is in a generous and boundless situation.

设计说明：
鉴于本案有临海特色，设计师以私人度假酒店及"海洋之心"为主题概念进行设计，将大海一日所呈现的各种宝石般的蓝绿色贯穿整个空间，并以象征阳光沙滩的白色作为陪衬，大气磅礴，一气呵成。

MATRIX DESIGN

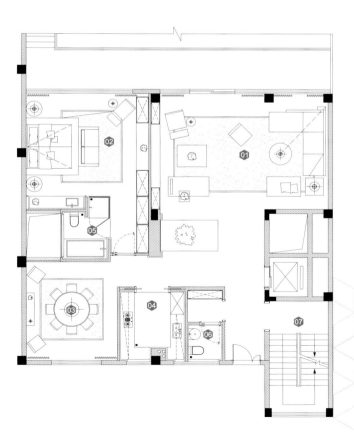

First Floor Plan 一层平面布置图
01.起居室
02.次卧
03.景观餐厅
04.厨房
05.次主卫
06.公卫
07.楼梯厅

MATRIX DESIGN

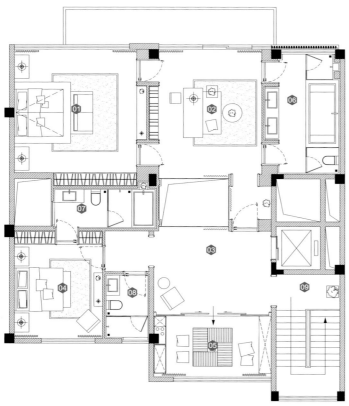

Second Floor Plan 二层平面布置图
01.主卧
02.起居书房
03.过廊
04.次卧
05.多功能家庭活动区
06.主卫
07.次卫
08.客卫
09.楼梯厅

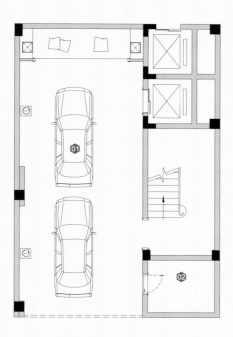

Ground Floor Plan 地下室平面布置图

01.景观车库大堂
02.设备房

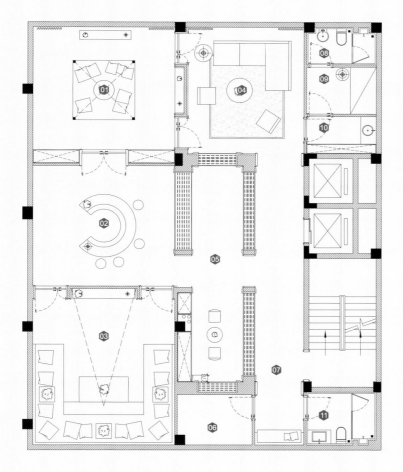

Ground Floor Plan 地下夹层平面布置图

01.特色多功能房、红酒雪茄房
02.酒吧
03.特色视听室
04.休闲室
05.水晶红酒房、雪茄房
06.设备房
07.艺术玄关
08.卫生间
09.工人房
10.工作室
11.洗手间

No. 7 Villa, Tian Yue Wan, Shenzhen Gemdale Real Estate

金地深圳天悦湾7号楼别墅空间

Project Contents: Villa Show Flat
Design Area: 600 m²
Hardware Decoration Cost: RMB 3500/m²
Soft Decoration Cost: RMB 3500/m²
Main Materials: Natural Stone, Mosaic, Solid Wood Flooring, Imported Carpets, Fur, White Brushing Lacquer, etc
Time of Design: Oct, 2010
Completion Time: May, 2011

项目性质：别墅样板房
设计面积：600平方米
硬装造价：3500元/平方米
软装造价：3500元/平方米
主要材料：天然石材、马赛克、实木地板、进口地毯、毛皮、白色手扫漆等
设计时间：2010年10月
竣工时间：2011年05月

Design Specification:

In this project, beige is matched with gold within the framework of French classic lines; from the whole to parts, the design is more like a diversified way of thinking, which combines the classical romantic feelings with modern people's needs towards life as well as integrates sumptuousness and elegance with fashion and modernity.

On the facade, the classical pure white line wallboard is matched with the wallpaper with modern patterns. The designer uses the stone parquet on the ground and adopts the crude texture and natural colors of stone to represent the sense of quality of the entire space, which makes the luxury, grade and taste of the living room and the bedroom flow without reservation. In terms of furniture collocation, the solid wood furniture made of wood and panels as well as the lacquered surface of furniture with a semi-closed effect not only can fully show the texture of veneer, but also make people feel the smoothness and tidiness behind the lacquered veneer when they touch it In the soft decoration, the designer adopts white, light yellow, dark brown and other colors as the keynote and blends a little gold to make the color look bright and generous as well as make the whole space give people an open, tolerant and extraordinary manner.

设计说明：

本案的设计在法式经典线条的骨架中，以米色搭配金色，从整体到局部，更像是一种多元化的思考方式，将古典的浪漫情怀与现代人对生活的需求相结合，兼容华贵典雅与时尚现代。

在立面上用古典、纯白色的线条墙板，搭配具有现代纹样的墙纸。地面则采用石材拼花，用石材天然的纹理和自然的色彩来体现整个空间的品质感，使客厅和卧室的奢华、档次和品位毫无保留地展现出来。在家具配置上，板木结合的实木家具，家具漆面具有半封闭漆效果，不仅能将木皮的纹理尽情展示，在徒手触摸时，还能感受到油漆饰面的光滑、平整。在软装配饰上使用白色、浅黄色和暗褐色作为基调，糅合少量金色，使色彩看起来更加明亮、大方，整个空间给人一种开放、宽容的非凡气度。

MATRIX DESIGN

MATRIX DESIGN

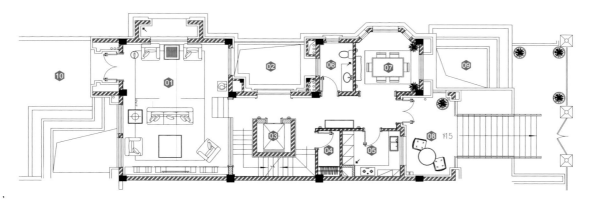

First Floor Plan 一层平面布置图
01.起居室
02.采光井
03.电梯
04.更衣室
05.厨房
06.卫生间
07.餐厅
08.入口平台
09.下沉花园
10.后院

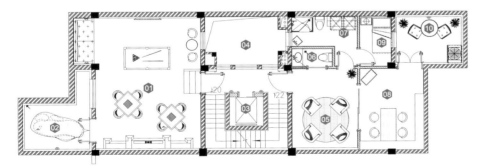

B1 Floor Plan 负一层平面布置图

01.休闲室
02.采光井
03.电梯
04.采光井
05.次卧室
06.卫生间
07.洗衣房
08.红酒房
09.工人房
10.后院

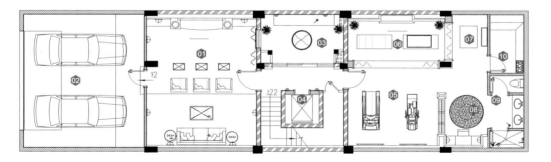

B2 Floor Plan 负二层平面布置图

01.家庭影院
02.车库
03.庭院
04.电梯
05.健身区
06.按摩区
07.更衣区
08.化妆间
09.卫生间
10.桑拿房

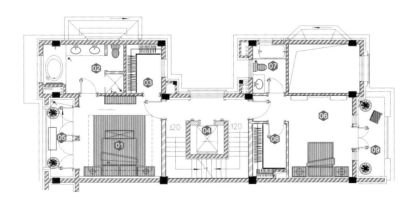

Second Floor Plan 二层平面布置图
01.二层主卧
02.主卫
03.主卧更衣室
04.电梯
05.阳台
06.二层客卧
07.客卧卫生间
08.客卧更衣室
09.阳台

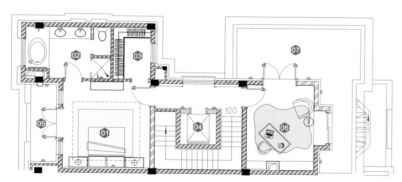

Third Floor Plan 三层平面布置图

01.三层主卧
02.主卫
03.主卧更衣室
04.电梯
05.阳台
06.书房
07.露台

MATRIX DESIGN

104/105

No. 3 Villa, Tian Yue Wan, Shenzhen Gemdale Real Estate

金地深圳天悦湾3号楼别墅空间

Project Contents: Villa Show Flat
Design Area: 350 m²
Hardware Decoration Cost: RMB 3500/m²
Soft Decoration Cost: RMB 3500/m²
Main Materials: Cherry Wood Veneer, Stone, Paint, Wallpaper, Wood Flooring, Mosaic, Culture Stone, etc.
Time of Design: Oct, 2010
Completion Time: May, 2011
Design Specification: The project is positioned as the American style and shows the grace and generousness as well as magnificence and luxury of space through the fashionable color expression, and at the same time the designer of the project takes the comfortable function as the guidance to emphasize nature and leisure. At the beginning of design, the designer firstly confirms the style, color and fabrics of the main furniture in order to determine the materials and colors of decoration. American furniture, solid wood floor, cloth sofa and dark color curtains make the color of the whole space elegant and simple. The perfect collocation of stone, wood and cloth is fully represented here and much increases the luxurious temperament. The large volume furniture meets the visual needs and in the meantime increases the comfortable feel. The accessories with classical elements are added to create the sense of yearning comfort and quietness.

项目性质：别墅样板房
设计面积：350平方米
硬装造价：3500元/平方米
软装造价：3500元/平方米
主要材料：樱桃木饰面、石材、涂料、墙纸、木地板、工艺马赛克、文化石等
设计时间：2010年10月
竣工时间：2011年05月
设计说明：本案定位为美式风格，通过时尚的色彩表现，彰显出空间的雍容大气、气派豪华，同时以舒适功能为导向，强调自然休闲。设计之初，设计师首先确定了主要家具的款式、色彩和面料，以此确定装修的材料及色彩。美式家具、实木地板、布艺沙发、深色窗帘，使空间整体色彩典雅质朴。石材、木材与布艺的完美搭配在此得到充分体现，更增添了奢华的气质。大体量的家具，在满足视觉需求的同时，也增强了舒适感。配饰加入古典元素，营造出一种令人向往的舒适宁静之感。

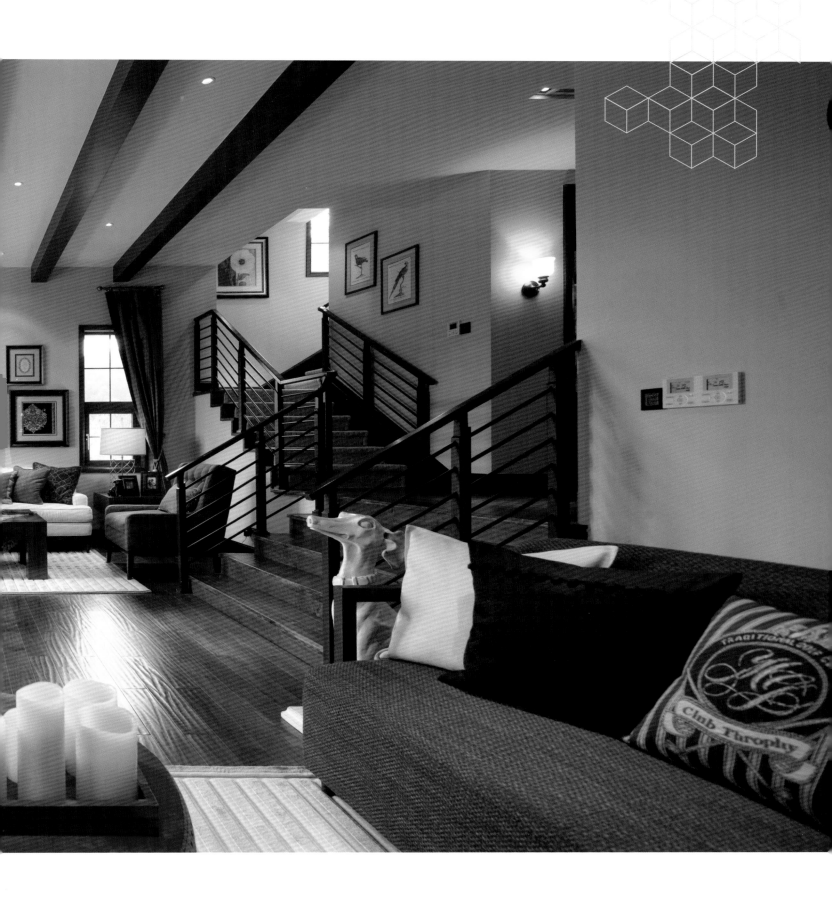

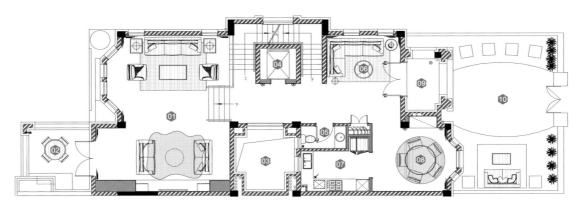

First Floor Plan 一层平面布置图
01.家庭室
02.阳台
03.采光井
04.电梯
05.卫生间
06.起居室
07.厨房间
08.餐厅
09.入户平台
10.前院

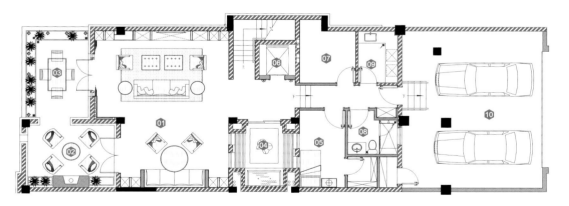

B1 Floor Plan 负一层平面布置图
01.休闲厅
02.室外休闲厅
03.庭院
04.内庭院
05.工人房
06.电梯
07.机房
08.卫生间
09.洗衣房
10.车库

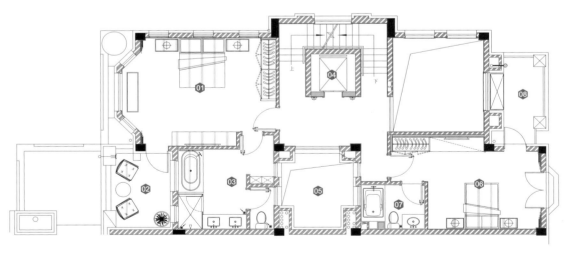

Second Floor Plan 二层平面布置图
01.二层主卧
02.阳台
03.主卫
04.电梯
05.采光井
06.二层客卧
07.客卫
08.阳台

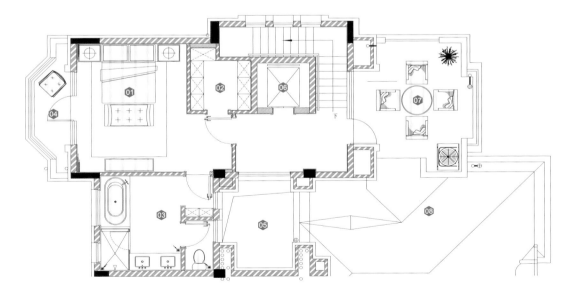

Third Floor Plan 三层平面布置图

01. 三层主卧
02. 主卧更衣室
03. 主卫
04. 阳台
05. 采光井
06. 电梯
07. 休闲阳台
08. 屋面

MATRIX DESIGN

116/117

Sales Center of Wonder Town, Vanke Real Estate, Chongqing

万科地产（重庆）缇香郡销售中心

Project Contents: Sales & Demonstration Area
Design Area: 586 m²
Hardware Decoration Cost: RMB 2000/m²
Soft Decoration Cost: RMB 2000/m²
Main Materials: Microporous Volcanoic Rock, Teak Stain, Quartersawn Oak
Time of Design: Mar, 2010
Completion Time: Oct, 2010

项目性质：销售展示
设计面积：586平方米
硬装造价：2000元/平方米
软装造价：2000元/平方米
主要材料：微孔火山岩、柚木染色、直纹橡木
设计时间：2010年03月
竣工时间：2010年10月

Design Specification:

The project is located in Chongqing, a beautiful mountain city, and the sales centre of Vanke Wonder Town is combined by several foreign-style houses by knocking the walls through to make it a whole exhibition space. With the existing space in disorder and low storey height, the function exhibition to fulfill the demand of sales becomes the key point of the design. Besides, the design and construction period is short and the cost is strictly controlled, the design makes full use of the specialty of the winding space of the original construction by means of view borrowing to combine one exhibition space with another. The central view area, made two-storey, is painted white by large area in sharp contrast with the dark floor and ceiling. The cultural elements with the oriental spirits are integrated into the exhibition area which is for sales.

设计说明：

本案位于美丽的山城重庆，是由几套洋房的一层打通合并形成的展示空间。现有的空间层次杂乱，层高较低，满足销售为目的的功能展示需求成为其设计的重点，同时兼顾很短的设计施工周期及严格的造价要求。设计利用了原建筑空间动线曲折的特点，以借景的手法将各个展示空间串联起来，并将其中心造景区以二层通高的方式处理，利用大量的留白与深色的地面天花进行对比。在以销售为目的的展示空间中，加入了带有东方气质的文化诉求。

MATRIX DESIGN

122/123

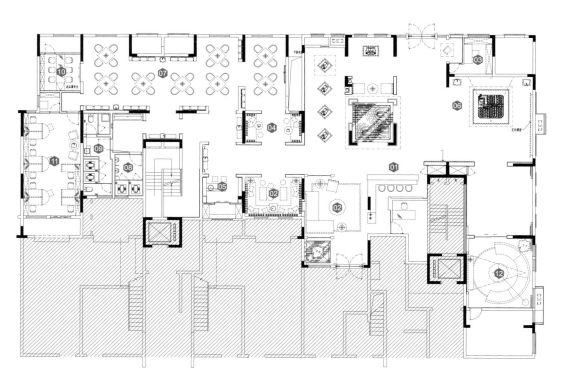

Ground Floor Plan 平面布置图
01.接待区
02.休息区
03.总控室
04.洽谈室
05.水吧服务区
06.展示区
07.洽谈区
08.男洗手间
09.女洗手间
10.财务室
11.地产及营销办公室
12.视听影音室

MATRIX DESIGN

MATRIX DESIGN

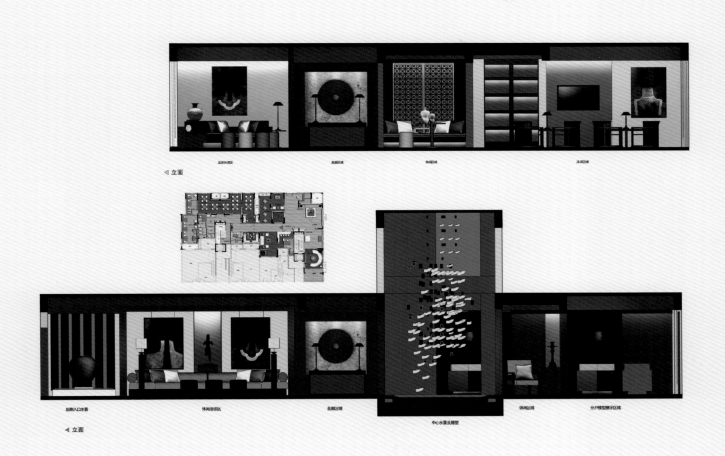

Show Flat D2, Joy Town, Wuxi

无锡复地悦城D2户型

Project Contents: Show Flat
Design Area: 120 m²
Hardware Decoration Cost: RMB 2500/m²
Soft Decoration Cost: RMB 2500/m²
Main Materials: Teak Dyeing, Textile Wallpaper, Sketch Wallpaper, Wood Flooring, White Travertine
Time of Design: May, 2012
Completion Time: Jul, 2012

项目性质：样板房
设计面积：120平方米
硬装造价：2500元/平方米
软装造价：2500元/平方米
主要材料：柚木染色、墙布、手绘墙纸、木地板、白洞石
设计时间：2012年03月
竣工时间：2012年07月

Design Specification:
Four-bedroom units of 120 square meters, compact and practical, this residential flat with ancient artistic charm set close to the Beijing-Hangzhou Grand Canal quietly into the Wuxi local Qinhuai landscape with our present life.

设计说明:
120平方米的四房户型,紧凑而又实用,设计时将古韵风雅加入了这套靠近京杭大运河的平层住宅之中,将无锡本地的秦淮风景与我们当下实际的生活悄然融入。

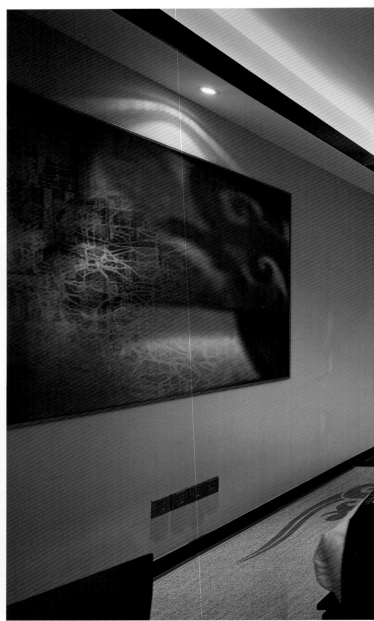

MATRIX DESIGN

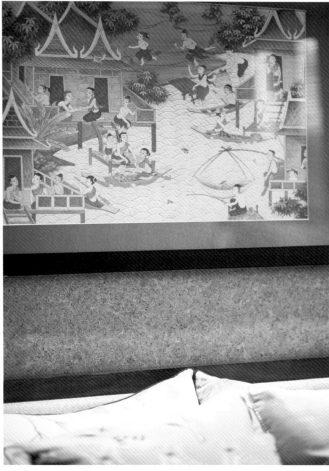

MATRIX DESIGN

Ground Floor Plan 平面布置图
01.客厅
02.餐厅/厨房
03.景观阳台
04.家庭厅
05.次卧室
06.主卧室
07.衣帽间
08.书房
09.主卫
10.公卫
11.生活阳台

Peng Cheng Office, Shenzhen

彭成实业有限公司办公室（深圳）

Project Contents: Office Space
Design Area: 1000 m²
Hardware Decoration Cost: RMB 2000/m²
Soft Decoration Cost: RMB 1000/m²
Main Materials: Wheat Straw Board, White Maple Veneer, Ultra Clear Glass, Mirror Finished Stainless Steel, etc.
Time of Design: Mar, 2011
Completion Time: Oct, 2011

项目性质：办公空间
设计面积：1000平方米
硬装造价：2000元/平方米
软装造价：1000元/平方米
主要材料：麦秸板、白枫木饰面、超白玻璃、镜面不锈钢等
设计时间：2011年03月
竣工时间：2011年10月

Design Specification:

The two-storey hollow reception hall is the key area of the entire space in the design, and the hollow background wall is the most important. The background wall takes the design concept of thriving, and then use wheat straw board and other renewable environmentally friendly materials to make a collage and superimposition so as to form a picture full of visual impact and artistic quality, meaning the company's thriving and flourish vista. Besides, it is also in line with the current design concepts of environmental protection and sustainable development. This project combines the modern minimalist style with a few New Oriental elements. Together with neat and pleasing light wood color, such a kind of design creates a positive, open, modern and international corporate image.

设计说明：
设计中两层高的中空接待大厅是整个空间的重点区域，其中空的背景墙是重中之重。背景墙运用"风生水起"的设计概念，运用麦秸板等可再生环保材料进行艺术拼贴与叠加，构成一幅极具视觉冲击力与艺术性的画面，寓意公司风生水起，蓬勃发展。另一方面也符合当前环保、可持续发展的绿色设计概念。将现代简约的设计风格与少许新东方的元素糅合一起，配以简洁、明快的浅木色，营造出一个积极开放、现代且国际化的公司形象。

MATRIX DESIGN

MATRIX DESIGN

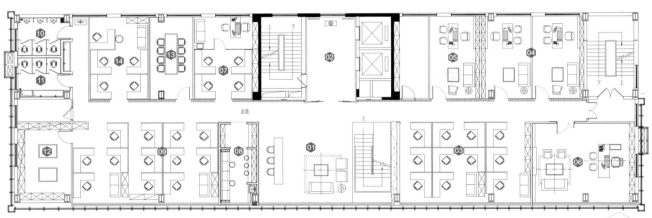

First Floor Plan 一层平面布置图
01.接待区
02.电梯厅
03.采购部/营销部
04.副总办公室
05.副总办公室
06.总经理办公室
07.行政人事部
08.文印、水吧区
09.成本部、工程部、设计部
10.男卫
11.女卫
12.材料样板间
13.中会议室
14.开发部

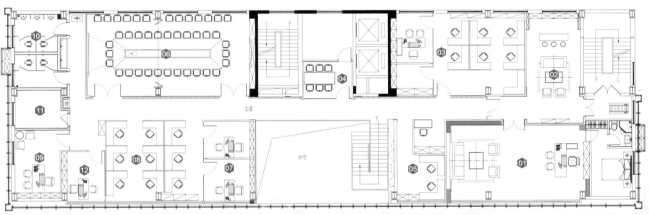

Second Floor Plan 二层平面布置图
01.董事长办公室
02.董事长会客厅
03.财务中心
04.小会议室
05.风控管理中心
06.大会议室
07.副总办公室
08.集团运营中心
09.总经理办公室
10.男卫
11.女卫
12.副总办公室

MATRIX DESIGN

APYRITE 碧玺

Project Contents: Show Flat
Design Area: 186 m²
Hardware Decoration Cost: RMB 2500/m²
Soft Decoration Cost: RMB 2000/m²
Main Materials: Ebony, Cloth Hard Package, Gold Foil, Ancient Wood Textured Stone, Bird of Paradise Stone, etc
Time of Design: Apr, 2010
Completion Time: Sep, 2010
Design Specification: How to explain the modern Chinese luxury style?Techniques of material contrast and abstraction of imageries are adopted to carry out the design in the project in DaDong Cheng with a building area of 184 m2 . Like the Chinese ink and wash painting from the ancient Chinese poets and literary men, the black serpenggiante is in contrast to the white stone pillars which look like the ancient scrolls and the background wall. The dark ebony stands for the long history and richness of Chinese culture. The golden edging, as the finshing touch highlights the delicate hand painted wall and the upholstery with typical patterns. On the other hand, the dark mirror material brings a dynamic space by reflection. The whole space is in an obvious contrast with outstanding effect, gorgeous, graceful, rich and humanistic.

项目性质：样板房
设计面积：186平方米
硬装造价：2500元/平方米
软装造价：2000元/平方米
主要材料：黑檀木、布艺硬包、金箔、古木纹石材、天堂鸟石材等
设计时间：2010年04月
竣工时间：2010年09月
设计说明：以中式为主题的现代奢华风格该如何演绎？在大东城的这套186平方米的豪宅设计中，设计师采用了材质语汇的对比及造型意象的提炼升华来对其进行设计处理。 黑色的木纹石犹如文人笔下的水墨映衬出以"书卷"为意象的白色石材柱及背景墙面，深色的黑檀木代表了深沉悠长的文化底蕴，金色的镶边作为点睛之笔，突出表现手绘的绢绣墙面及有代表性纹样的软包。黑色镜面材质的反射效果为空间增添了变化。 整个空间对比强烈、效果突出，华丽不失儒雅、厚重不乏人文。

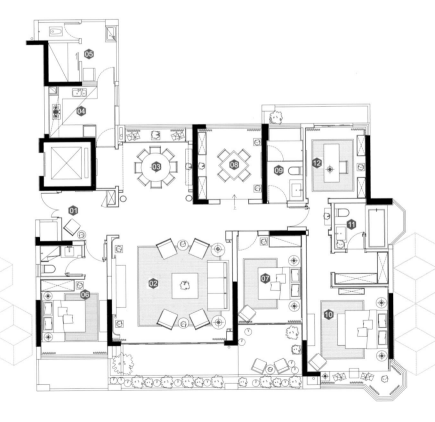

Ground Floor Plan 平面布置图
01. 门厅
02. 客厅
03. 餐厅
04. 厨房
05. 工人房
06. 老人房
07. 儿童房
08. 多功能房
09. 公共洗手间
10. 主卧室
11. 主卧卫生间
12. 书房

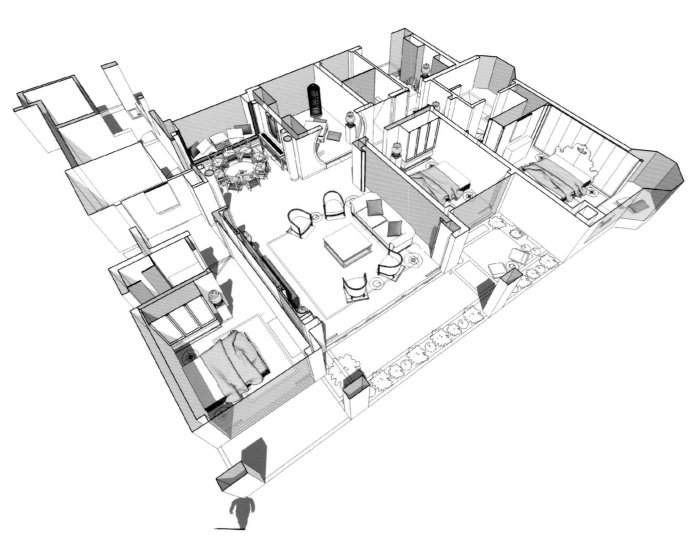

MATRIX DESIGN

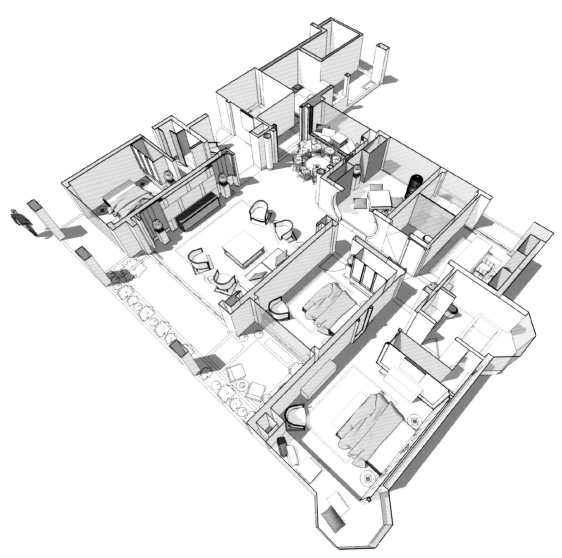

MATRIX DESIGN

Left Bank of Bordeaux
波尔多左岸

The designer expects to deliver a unique life of grace and luxury through the joint-row villa (townhouse), where the project area is 400 m² and living facilities are all available. The whole space uses rich colors of sandy beige, almond, grey beige, brown and cream white against colors of golden, silver, and soul colors of wine red, carmine and gray purple.
The wide use of imported marble, wallpaper, golden soft fabric makes the high quality and luxury stand out. Furniture and accessories with the classic elements bring us pleasant affinities and humanistic atmosphere. The whole design for Left Bank of Bordeaux is special, meaningful and thought-provoking, just like the famous red wine from Bordeaux.

对于一栋400平方米的联排别墅来说，应有的各项功能配置都一应俱全，设计师希望在这里展现一种别致优雅的奢华生活。整个空间色调丰富，以驼色、杏仁色、米灰色、褐色、乳白色来映衬金色、银色及点睛的酒红色、洋红色、灰紫色。大面积的进口云石、墙纸、烫金软包面料的运用突出了整个空间的奢华感和高档质量。
经典元素的家私和配饰的运用更带来了极具亲和力的人文感受，犹如那著名的波尔多左岸盛产的红酒，香醇浓郁、悠远流长。

MATRIX DESIGN

MATRIX DESIGN

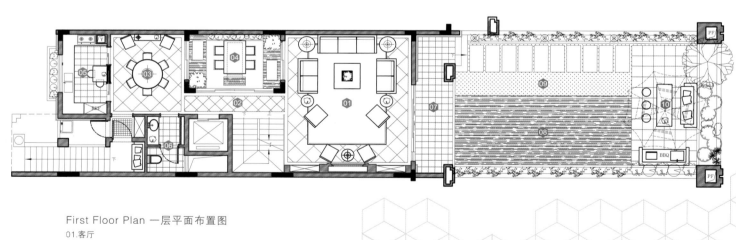

First Floor Plan 一层平面布置图
01.客厅
02.过廊
03.餐厅
04.阳光流水餐厅
05.厨房
06.客卫
07.露台
08.景观池
09.草坪
10.家庭聚会空间

Second Floor Plan 二层平面布置图
01.主卧
02.书房
03.玄关
04.衣帽间
05.主卫
06.过廊
07.儿卧
08.洗手间
09.休闲露台/高尔夫推杆联系区
10.露台

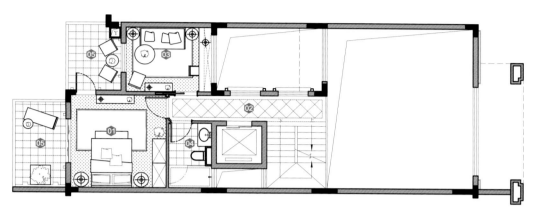

Third Floor Plan 三层平面布置图
01.次卧
02.过廊
03.多功能亲子房/家庭厅、客房
04.洗手间
05.景观露台

MATRIX DESIGN

MATRIX DESIGN

B1 Floor Plan 负一层平面布置图
01. 视听娱乐区
02. 过廊
03. 红酒雪茄吧
04. 桌球活动室
05. 工人房
06. 洗手间
07. 玄关
08. 车库

Hair Q Salon ,HK
香港Hair Q Salon 形象中心
(陈列宝个人作品)

Project Contents: Retail
Design Area: 70 m²
Hardware Decoration Cost: RMB 3000/m²
Soft Decoration Cost: RMB 3000/m²
Main Materials: Yellow Paint, GRG. Stainless Steel, etc.
Time of Design: Jun, 2009
Completion Time: Dec, 2009

项目性质：零售
设计面积：70平方米
硬装造价：3000元/平方米
软装造价：3000元/平方米
主要材料：黄色烤漆面板造型定制等
设计时间：2009年06月
竣工时间：2009年12月

Design Specification:
Hair Q Salon is in Hongkong Tuen Mun District Chelsea Square commercial space. Its design is followed by Lamborghini design concept - making use of the color of yellow and the folding type structure to create a futuristic fashion commercial space.
The designer hopes to combine all the functional demands with the folding type and artistic modeling. In this way, it achieves a high fusion of people's needs and designer's inspiration.

设计说明:
Hair Q Salon位于香港屯门区切尔西广场商业空间。设计以兰博基尼为设计概念,以黄色与折叠构造融合空间,打造具有未来感的时尚商业空间。
设计师希望通过一个折叠且富有艺术感的造型穿插到室内空间去,并将所有功能性的需要结合到该造型中去。通过这样的方式,项目的设计使功能性与设计师的设计灵感达到高度融合。

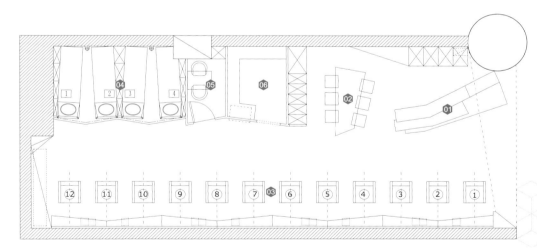

Ground Floor Plan 平面布置图
01.服务台
02.洽谈区
03.剪发、烫染位
04.洗发区
05.经理室
06.员工休息区

Switzerland Royal Noble Anti-Ageing Company Reception Club, ShenZhen

SRN瑞士皇家抗衰老公司（深圳接待会所）

(陈列宝个人作品)

Project Contents: Club Space
Design Area: 200 m²
Hardware Decoration Cost: RMB 3000/m²
Soft Decoration Cost: RMB 3000/m²
Main Materials: Red Travertine, GRG, Automotive Enamel, Mirror Finish Stainless Steel, etc.
Time of Design: Jun, 2011
Completion Time: Oct, 2011

项目性质：会所空间
设计面积：200平方米
硬装造价：3000元/平方米
软装造价：3000元/平方米
主要材料：红洞石，高强度玻璃纤维树脂，汽车漆，镜面不锈钢等
设计时间：2011年06月
竣工时间：2011年10月

MATRIX DESIGN

Design Specification

The club will be used as the reception hall for SRN's representative office in China. Designer hopes to seek for design innovation and breakthrough on the basis of Switzerland royal temperament. Through use of simple European design style, coupled with modern structured sofa and lamp posts, and the luxury European furniture, designer makes the design between modern and classical. Using red and white Swiss national flag as the main colors, it can not only show the temperament of the owner, but also give the silent emphasis of Switzerland royalty to visitors.

Designer blends various elements with strong point and practical functionalities into an organic whole to create SRN's corporate image, which is born in Switzerland wealthy, and facing future. It offers a luxury and comfortable rest space and provides a contact window for visitors to get to know more about the historical beauty agency.

设计说明：

该项目作为SRN瑞士皇家抗衰老公司在中国办事处机构的接待会所，设计师希望能够在保证瑞士皇室的气质上，寻求创新与突破。通过运用简欧的设计风格，加上具有现代结构感的沙发与灯柱设计，再配以厚重感的豪华欧式家具，使设计在现代与古典中游离。色彩上采用瑞士国旗的红白两色作为设计的主色调，不但能章显业主本身气质，对于访客来说也是对该公司来自瑞士的一种无声的强调。

设计师将各种带有强烈指向性的元素与实用性功能有机的糅合成一个整体，打造出一个源于瑞士、出生豪门、面向未来的SRN公司形象，不但提供了一个豪华舒适的接待空间，也让访客最大程度地了解这个具有百年悠久历史的美容抗衰老机构。

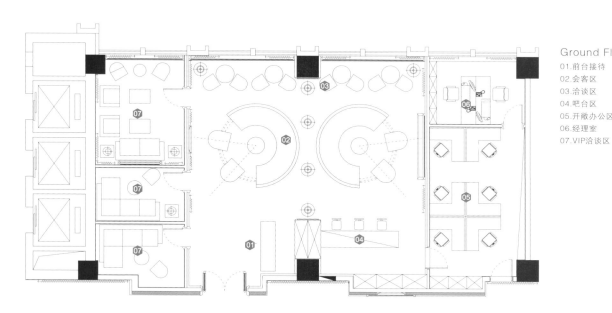

Ground Floor Plan 平面布置图
01.前台接待
02.会客区
03.洽谈区
04.吧台区
05.开敞办公区
06.经理室
07.VIP洽谈区

Show Flat, Uptown, Chongqing

重庆复地上城示范单位

(王冠个人作品)

Project Contents: Show Flat
Design Area: 70 m²
Hardware Decoration Cost: RMB 2000/m²
Soft Decoration Cost: RMB 2500/m²
Main Materials: Ceramic Tile, Mosaic, Leather, Ebony, etc.
Time of Design: Feb, 2009
Completion Time: Oct, 2009

项目性质：样板房
设计面积：70平方米
硬装造价：2000元/平方米
软装造价：2500元/平方米
主要材料：瓷砖、马赛克、皮革、黑檀木等
设计时间：2009年02月
竣工时间：2009年10月

Design Specification:
This project locates right in the New District of north Chongqing city. It is absolutely an exciting real estate project which is full of vitality. 70 m^2 of two-bedroom apartment, with the interior wall knocked through, is now becoming one open-space, which only contains clear communal area and rest area. "Lyric style combined with album-rock in black and white" is chosen to be the theme of project, meeting the yearnings of young generation but at the same time it also superbly expresses their own personalities.

设计说明：
本案位于重庆北部新区，是一个充满生命力的地产项目。70平方米的两室空间被打通，形成一个开敞的大空间，仅分为公共活动区和休闲睡眠区。以黑白摇滚的抒情风格作为设计主题，以迎合年青人对品质生活的向往与追求，同时又将自我的个性彰显得淋漓尽致。

MATRIX DESIGN

MATRIX DESIGN

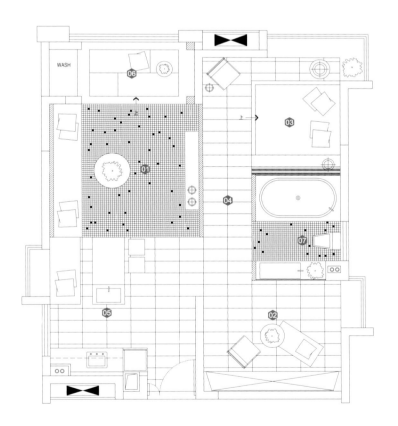

Ground Floor Plan 平面布置图
01.起居区
02.工作休息区
03.睡眠区
04.过道
05.餐吧区
06.游戏区
07.洗手间

Joy Town, Shenzhen

深圳中央悦城别墅

(王冠个人作品)

Project Contents: Villa Show Flat
Design Area: 580 m²
Hardware Decoration Cost: RMB 2000/m²
Soft Decoration Cost: RMB 2000/m²
Main Materials: White Travertine, Old Wood Flooring, Red Glass, Gold Foil
Time of Design: Sep, 2006
Completion Time: Dec, 2006
Design Specification: The greatest feature of this project lies in the rich colors and lines of varied shapes. The designer develops the use of the colors in the living room to the extreme – white of the sofa, black of the table lamp cover, copper yellow on the coffee table, green black of the pillows, dark red of the floor, light red of the window screens... dyes all colorful throughout the whole living room. In addition, the bold use of red also highlights the great originality performance of the design. The red curtain shows light of the elegant in the breeze. The red walls encourage people and stimulate people's enthusiasm and aspirations for life.

项目性质： 别墅样板房
设计面积： 580平方米
硬装造价： 2000元/平方米
软装造价： 2000元/平方米
主要材料： 白洞石材、老木头地板、红色玻璃、金箔
设计时间： 2006年09月
竣工时间： 2006年12月
设计说明：本案最大的特点在于色彩的丰富和线条造型的多变。设计师将客厅里的色彩运用发挥到了极致。沙发的白、台灯罩的黑、茶几饰品的铜黄、抱枕的墨绿、地板的深红、窗纱的浅红……将整个客厅渲染得五彩斑斓。此外，红色的大胆运用也突出地表现了设计的别具匠心。红色的窗幔，在清风的轻拂下展现出轻盈的飘逸之态；红色的墙壁，还能激发人们对生活的热情与渴望，催人奋发向上。

MATRIX DESIGN

MATRIX DESIGN

204/205

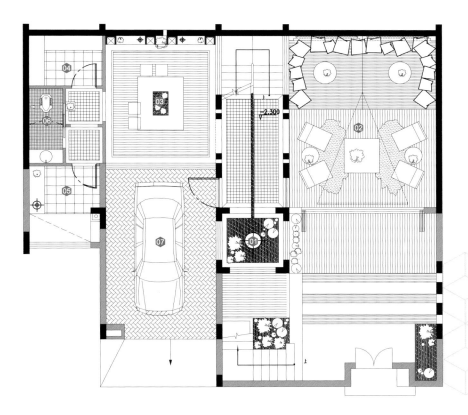

B1 Floor Plan 负一层平面布置图
01.玄关
02.家庭娱乐厅
03.情景酒吧间
04.工作间
05.工人房
06.卫生间
07.车库

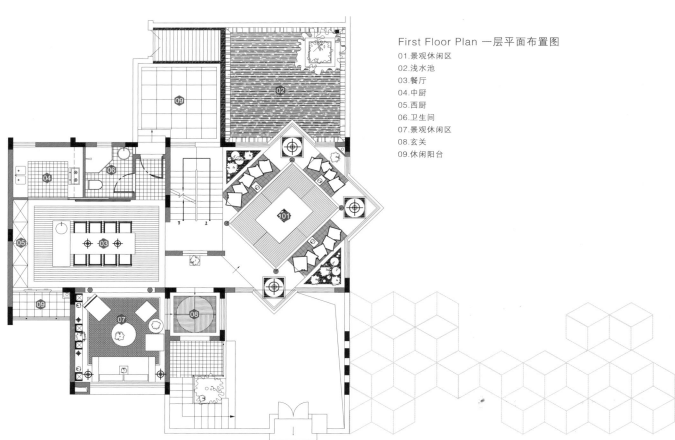

First Floor Plan 一层平面布置图
01.景观休闲区
02.浅水池
03.餐厅
04.中厨
05.西厨
06.卫生间
07.景观休闲区
08.玄关
09.休闲阳台

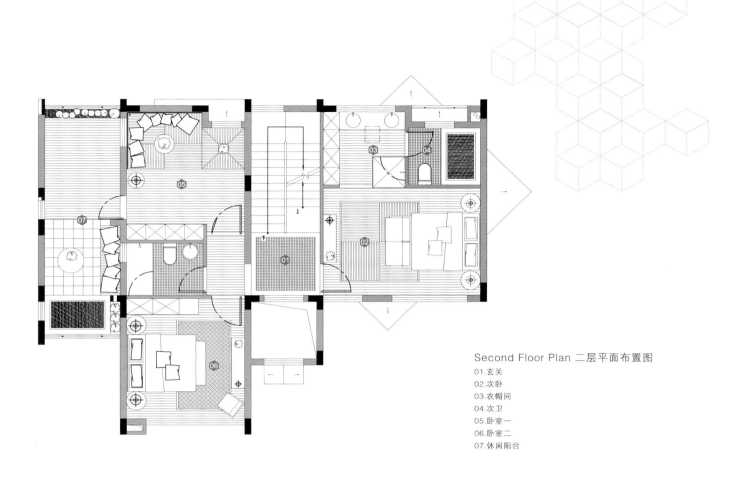

Second Floor Plan 二层平面布置图
01.玄关
02.次卧
03.衣帽间
04.次卫
05.卧室一
06.卧室二
07.休闲阳台

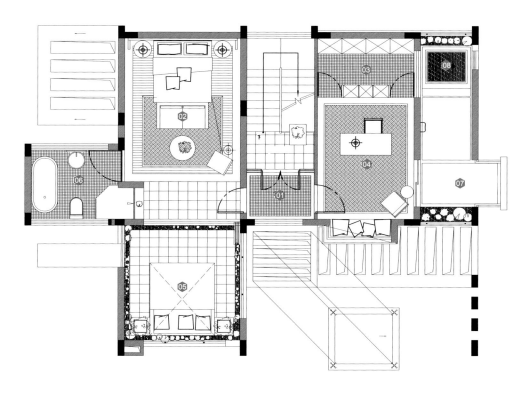

Third Floor Plan 三层平面布置图
01.玄关
02.主卧
03.衣帽间
04.书房
05.瑜伽及阳光午休区
06.主卫
07.景观阳台
08.SPA池

Penthouse, St-Maurice Garden, Shenzhen

圣莫丽斯顶层行宫

Project Contents: Villa Show Flat
Design Area: 240 m²
Hardware Decoration Cost: RMB 3000/m²
Soft Decoration Cost: RMB 2500/m²
Main Materials: Stone, Leather Hard Package, Imported wallpaper, Black Ebony Veneer, etc
Time of Design: Feb, 2011
Completion Time: Jul, 2011

项目性质：别墅样板房
设计面积：240平方米
硬装造价：3000元/平方米
软装造价：2500元/平方米
主要材料：石材、皮革硬包、进口墙纸、黑檀木饰面等
设计时间：2011年2月
竣工时间：2011年7月

Design Specification:

This is a very well-known house property in Shenzhen, the developer promotes its top-floor multiple penthouse at the end of the sale in order to highlight the sale. Therefore, the owner hopes to make it become a unique top palace because the product is scarce and can not be copied. So the designer changes the original cramped space into a separate staircase hall, and the private club concept encourages the designer to make full use of the recreational function of this open and wide space. What's more, the leisure recreation area (such as the red wine room, the audio-visual room) which only appears in the villa is now given to a multiple space which is originally a simple flat layer. So, it's the highlight of the sale!

设计说明：

这是一个在深圳很有知名度的楼盘，在销售接近尾声的时候开发商推出珍藏的顶层复式单位，由于产品的稀缺和不可复制性，业主希望将其打造成独一无二的顶层行宫。于是设计师调整了原来狭小拥挤的立体交通动线，开辟出了独立的楼梯厅，并将空中私享会所的概念引入其中，呼应了顶层露台开阔空间的娱乐功能。将一般只出现在别墅里的休闲娱乐区域（如红酒房、视听间等）赋予一个原本只是简单平层叠加形成的复式空间，为其销售加足了分！

First Floor Plan 一层平面布置图
01.客厅
02.餐厅
03.楼梯厅
04.书房/多功能房
05.次卧
06.次卫
07.老人房
08.公卫
09.厨房
10.工作间
11.休闲露台
12.客厅阳台

Second Floor Plan 二层平面布置图

01.影音室
02.水吧区
03.红酒房
04.楼梯厅
05.主卧
06.主卫
07.衣帽间
08.子女房
09.客卫
10.书房阳台
11.休闲露台

Third Floor Plan 三层平面布置图
01.BBQ家庭聚会区
02.瑜伽&阳光午休区
03.高尔夫推杆
04.SPA
05.水景

J Club, Verakin New Park City

同景地产国际城J会所

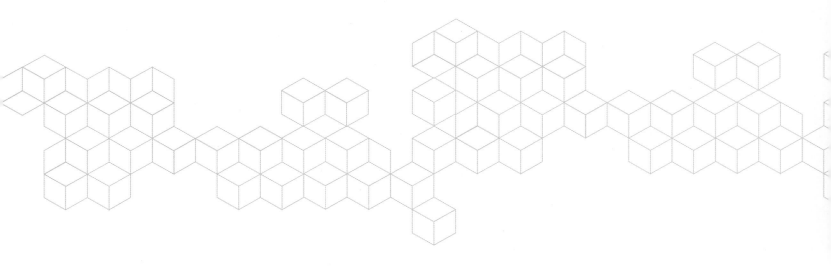

同景国际城J组团会所室内空间剖面分析

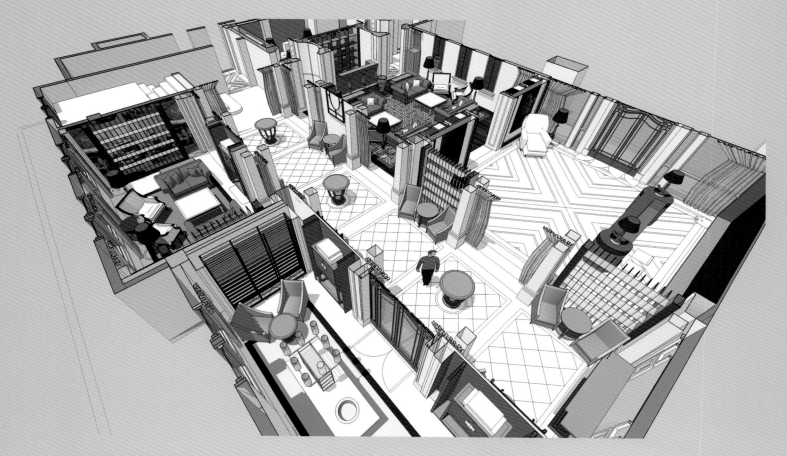

同景国际城J组团会所室内空间分析鸟瞰图

MATRIX DESIGN

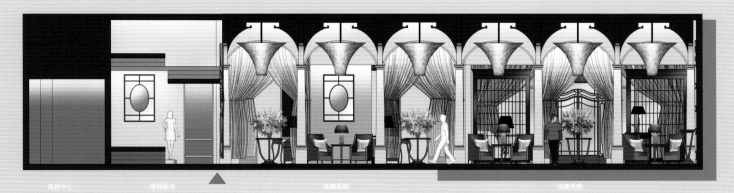

同景国际城J组团会所室内空间剖面分析

同景国际城J组团会所平面布置图

Office of Vanke Real Estate, Chongqing

万科地产（重庆）办公楼

Project Contents: Office Space
Design Area: 2000 m²
Hardware Decoration Cost: RMB 1800/m²
Soft Decoration Cost: RMB 1000/m²
Main Materials: Aluminum Profile, Ancient Wood Textured Marble, Iron (Fluorocarbon Paint), Wheat Straw Board, Cork, Bamboo Veneer, etc.
Time of Design: Apr, 2010
Completion Time: Dec, 2010

项目性质：办公室
设计面积：2000平方米
硬装造价：1800元/平方米
软装造价：1000元/平方米
主要材料：铝型材、古木纹大理石、黑铁（氟碳漆面）、麦秸板、软木、竹饰面等
设计时间：2010年04月
竣工时间：2010年12月

MATRIX DESIGN

Design Specification:

The design is closely related to the building's owner. This project belongs to Vanke Real Estate (Chongqing) Co., Ltd., so it adheres to Vanke's corporate culture which has always been advocated. At the beginning of the project, environment protection, energy saving, pollution-free, comfort, and building features are the key words of the starting point in the design. Therefore, in the selection of materials and construction process, the designers use a number of renewable and environmentally friendly materials, such as wheat straw board, cork, bamboo veneer. While the main backdrop wall adopts the used windows and doors made of aluminum profile and industrial aluminium sections, which not only saves the cost of materials, but also achieves an unexpected visual effects.

After the repeated "collisions" with the owner who is very professional in design, the space finally gets a lot of bright spots in this project, for example, the main backdrop wall shows the nature of the real estate and the local topography in Chongqing. Now this wall has become an iconic symbol of the building or even Vanke (Chongqing). The terraced staff recreation area, located in the center of the first floor, is known as a multi-zone with the function of reading, meeting, training, leisure, and brainstorming, and is also the best place to rest. The owner speaks highly of it. The remote video conference room in the second floor is a space full of sense of construction frame. Its design concept comes from the "Ark" concept, since the storey height allow the designers to do the design overhead the bottom. It is a very pleasant thing to discuss in such a conference room, so the room has the highest utilization rate in the entire building.

设计说明：

办公楼的设计和业主息息相关，本案业主为万科地产（重庆）有限公司，因此设计以万科一贯倡导的企业文化为指导思想，在项目之初的沟通中，环保节能、无污染、舒适、建筑感等的关键词成为设计的出发点，在选材和施工工艺上尽量选一些可再生及环保的材料作为主材，比如麦秸板、软木、竹饰面等都为速成材料，主背景墙的材料为废旧的铝型材门窗和工业铝型材的断面，既降低了材料的造价，也做出了意想不到的视觉效果。

在和专业度极高的业主反复"碰撞"之下产生了许多的"火花"亮点：主背景墙，既能反映地产的行业性质，又和重庆本土地貌结合在一起，现在已成为整个办公楼乃至整个重庆万科的标志性符号；一层中心阶梯式的员工休闲区，也可以称之为多功能区域，集阅读、开会、培训、休闲、头脑风暴室等功能于一体，是工作之余最好的休息场所，一得到业主的一致认可；二层的中心远程视频会议室，是一个建筑构架感十足的空间，设计之初理念来自于"方舟"概念，由于层高的允许，设计师将底部架空，在这样具有形式感的会议室里开会讨论，是件非常愉快的事，于是该会议室也成为整个办公楼使用率最高的会议室。

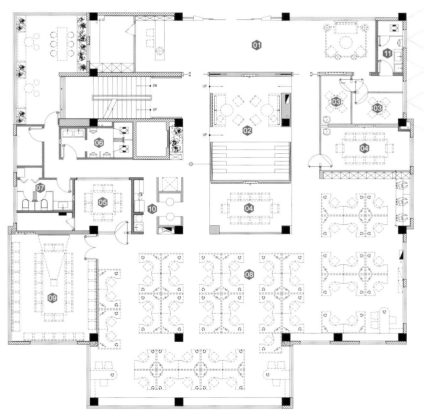

First Floor Plan 一层平面布置图
01.前厅接待
02.会客休闲区
03.洽谈室
04.小会议室
05.小会议室二
06.男卫
07.女卫
08.开放办公区
09.大会议室
10.文印室
11.客人卫

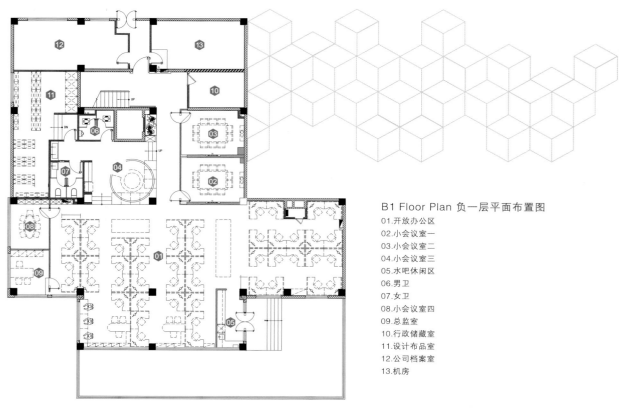

B1 Floor Plan 负一层平面布置图
01.开放办公区
02.小会议室一
03.小会议室二
04.小会议室三
05.水吧休闲区
06.男卫
07.女卫
08.小会议室四
09.总监室
10.行政储藏室
11.设计布品室
12.公司档案室
13.机房

Second Floor Plan 二层平面布置图
01.开放办公区
02.会谈区
03.小会议室一
04.小会议室二
05.接待室
06.总经理办公室
07.秘书办公室
08.财务办公区
09.男卫
10.女卫
11.财务档案室
12.副总办公室
13.助总办公室
14.阳台

MATRIX DESIGN

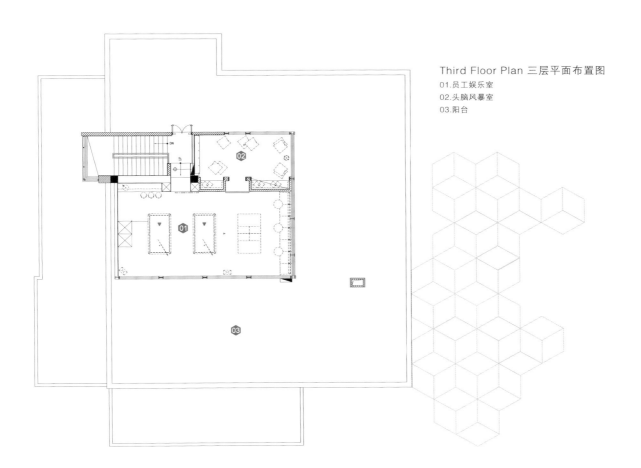

Third Floor Plan 三层平面布置图
01.员工娱乐室
02.头脑风暴室
03.阳台

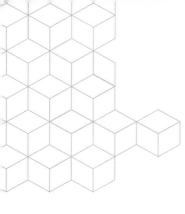

Vanke Health Club House
万科地产（重庆）天琴湾次会所

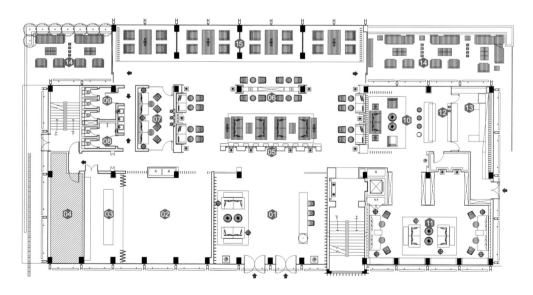

First Floor Plan 一层平面布置图
01.接待大堂
02.多功能区
03.厨房操作展示区
04.中餐厨房
05.过厅走廊
06.大堂休闲吧
07.雪茄吧
08.男洗手间
09.女洗手间
10.休闲阅读区
11.休闲阅读安静区
12.水吧吧台
13.简易操作间
14.户外休闲区
15.卡座休闲区

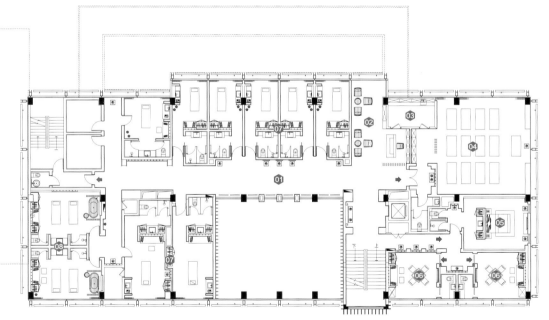

Second Floor Plan 二层平面布置图
01.过厅走廊
02.喝茶休息区
03.后勤储存
04.瑜伽教室
05.客房
06.棋牌室
07.SPA
08.SPA双人

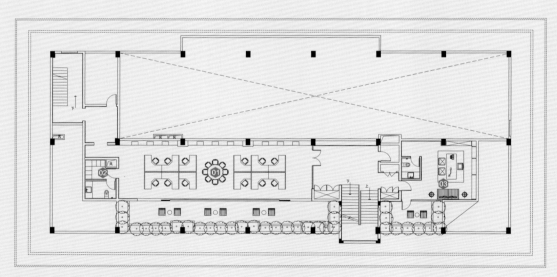

Third Floor Plan 三层平面布置图
01.办公区
02.洗手间
03.经理室

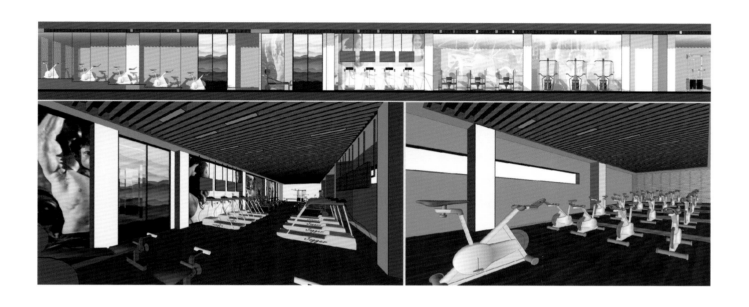

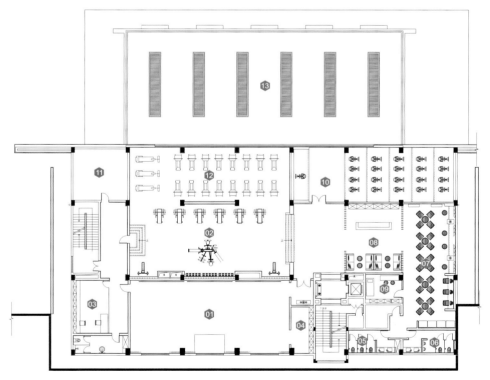

B1 Floor Plan 负一层平面布置图
01.有氧操房
02.组合器械区
03.教练室
04.操房换鞋室
05.女洗手间
06.男洗手间
07.休息区
08.接待及水吧区
09.体测及医务室
10.飞轮教室
11.储藏室
12.有氧训练区
13.游泳池屋顶

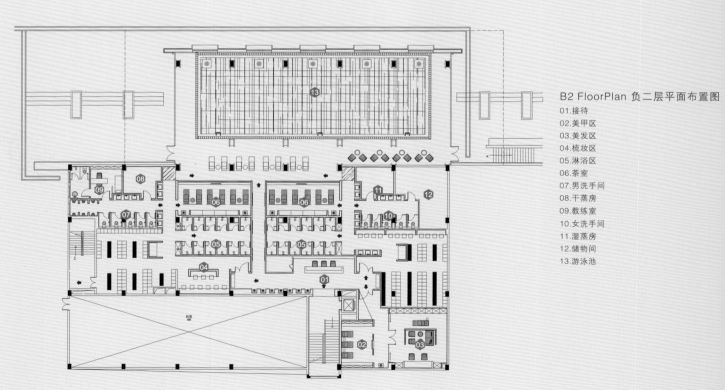

B2 FloorPlan 负二层平面布置图
01.接待
02.美甲区
03.美发区
04.梳妆区
05.淋浴区
06.茶室
07.男洗手间
08.干蒸房
09.教练室
10.女洗手间
11.湿蒸房
12.储物间
13.游泳池

MATRIX DESIGN

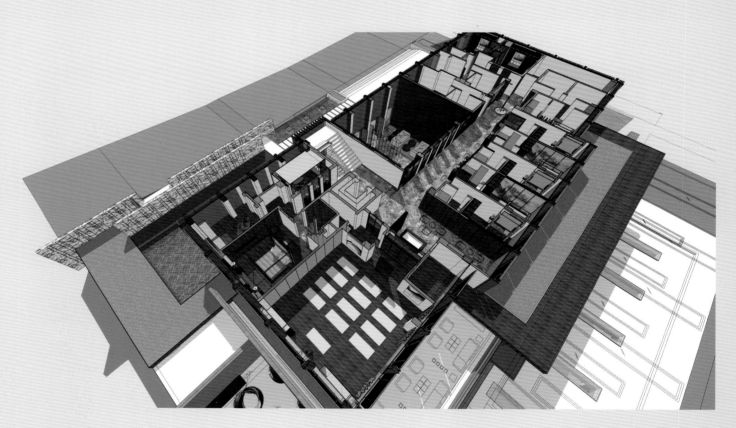

MATRIX DESIGN

Tian Qin Wan Golf Club, Vanke Real Estate, Chongqing

万科地产(重庆)天琴湾高尔夫会所

Project Contents: Golf Club, Sales & Demonstration Area
Design Area: 4000 m²
Hardware Decoration Cost: RMB 2000/m²
Soft Decoration Cost: RMB 2000/m²
Main Materials: Bawanghua Grantie, Big-Hole Volcanic Rock, White Sandstone, Dark Grey Hemp Stone, Fabric Metope Hard Roll, Multi Layer Parquet, Iron (Fluorocarbon Paint), Elm Wood Veneer, Gray Ceramic Tile, etc.
Time of Design: Apr, 2011
Completion Time: Dec, 2011
Design Specification: The project is full of challenges to the designer, but the designer gets the balance point when selecting between the various styles. In both the floor planning and the soft decoration, the designer focuses on the aim that the Japanese Simple Style of the order space to be shown, that the Southeast Asian Style details with careful sculptures to be reflected, that the beauty in form of the Chinese grand symmetrical style to be expressed. The outstanding shining point of the design especially reflected in the careful and cautious selections of furniture and furnishings. A perfect project is done with favorable climate, geographical position and support of the people. The most commendable thing is the owner's cooperation and contribution with no condition, which is one of the most important things, which finally leads to the success of the project.

项目性质: 高尔夫会所及销售展示
设计面积: 4000平方米
硬装造价: 2000元/平方米
软装造价: 2000元/平方米
主要材料: 霸王花花岗岩、大孔火山岩、白砂岩、深灰麻石、墙面布艺硬包、多层实木复合地板、黑铁(氟碳漆面)、榆木木饰面、灰色瓷砖等
设计时间: 2011年04月
竣工时间: 2011年10月
设计说明: 该项目对于设计师来说具有极高的挑战性,设计师在对各个风格的筛选与取舍之间找到了一个最佳的平衡点。在前期的平面规划及软装配饰中也遵循了这个宗旨:日式的简约矩阵的空间次序感、东南亚的精雕细琢的细节体现、中式大气轴对称的形式美,尤其在家具陈设艺术品的考究选择,是整个项目的亮点所在。一个好项目的完成,不仅是天时地利人和,更为难得的是甲方无条件的配合与付出,这个是至关重要的条件。

MATRIX DESIGN

260/261

268/269

MATRIX DESIGN

MATRIX DESIGN

MATRIX DESIGN

MATRIX DESIGN

284/285

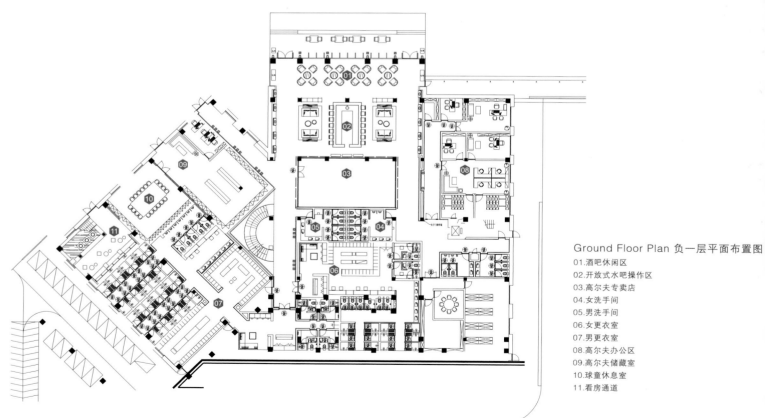

Ground Floor Plan 负一层平面布置图
01.酒吧休闲区
02.开放式水吧操作区
03.高尔夫专卖店
04.女洗手间
05.男洗手间
06.女更衣室
07.男更衣室
08.高尔夫办公区
09.高尔夫储藏室
10.球童休息室
11.看房通道

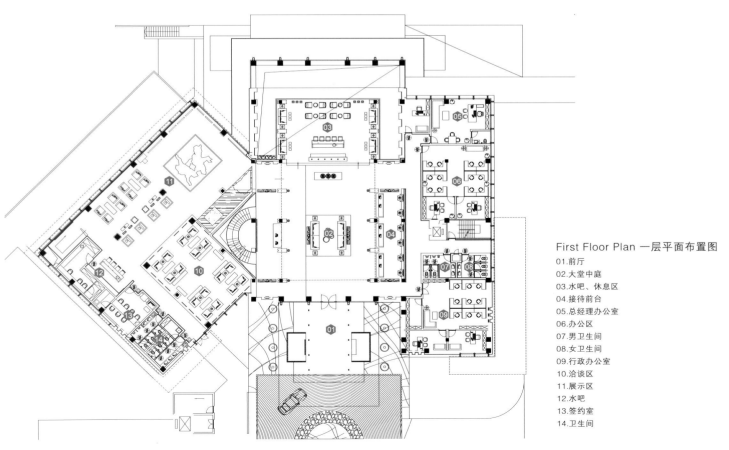

First Floor Plan 一层平面布置图
01.前厅
02.大堂中庭
03.水吧、休息区
04.接待前台
05.总经理办公室
06.办公区
07.男卫生间
08.女卫生间
09.行政办公室
10.洽谈区
11.展示区
12.水吧
13.签约室
14.卫生间

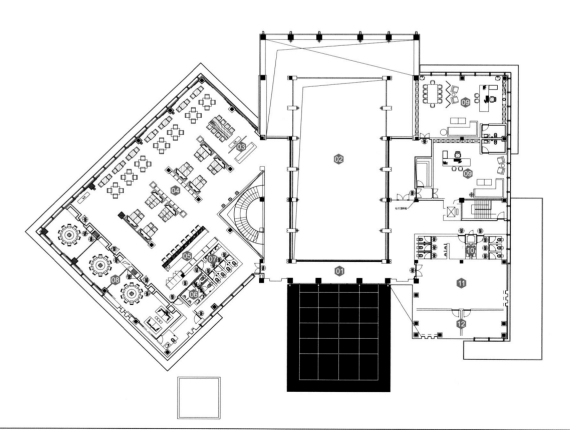

Second Floor Plan 二层平面布置图
01.回廊
02.大堂中庭上空
03.接待前台
04.餐厅
05.酒水吧
06.男卫
07.女卫
08.包房
09.董事长办公室
10.员工卫生间
11.厨房
12.储藏室

MATRIX DESIGN

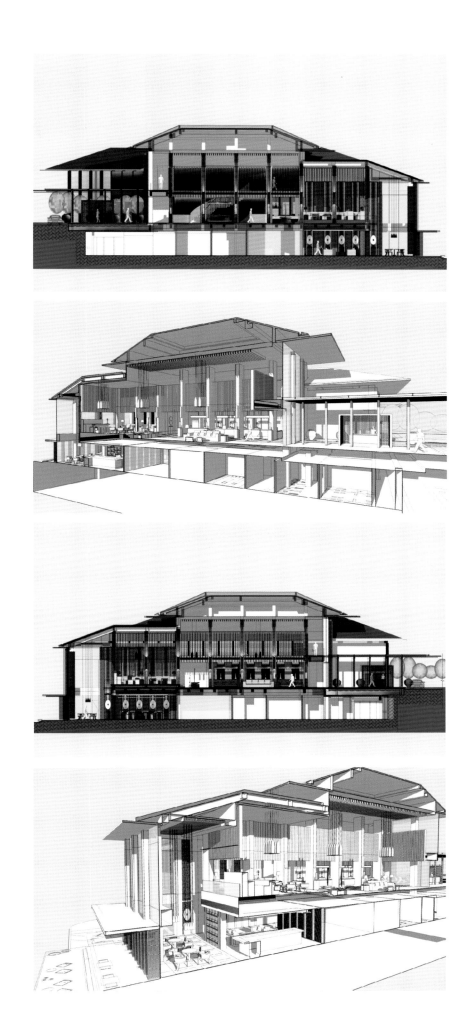

FRAME China Special Interview

While designing, treat every project, every first party and everyone of the users in the space with heart and soul and love.

用心、用情去对待每一个项目，每一个甲方，每一个在空间里面的使用者。

《FRAME国际中文版》，《MARK国际建筑设计》，《设计管理》杂志社就Matrix获得2012-2013年Andrew Martin国际室内设计大奖专访。

(F): Can you talk about the design concept of the award-winning project Main Club House of Vegavilla, Vanke Real Estate, Chongqing? What's the starting point of the design concept?
(M): The times now is full of beige and golden and we are all expecting some fresh "air". The world is full of foreign culture and we are all expecting the return of Oriental Culture. "New-Asian Style" is the strong power that we injected in our design, which is also the starting point of the design concept of the project. Comparing with the broad feeling in outer space, the inner space is full of magnificent auras. We focus on the special expression of the materials in a superficiality-off and decoration-off way.

(F): What's the secret of the design for this project? What made it different from the ordinary commercial space design?
(M): First of all, it is a sales center. However, we've reached a consensus with our client that we should focus on the relationship between the human being and the space instead of the sales relationship, that is to say, we should also emphasize on the cultural and casual auras while we focus on the sense of identity. If there is a secret in design, we consider that, we should treat every project, every first party and everyone of the users in the space with heart and soul and love.

(F): Did you use any special material or design method in this project?
(M): We used ordinary materials which represented regional characteristics in hard decoration to reduce the cost at the most, but we used the materials in soft decoration which reflected the South Asian Style, like shells and coconut shells etc. We made it in different permutation and combination ways to reach different effects. By exaggerate comparison in the sizes and the special treatment on the surface of the materials, the special texture of the materials can be shown clearly no matter under sunlight or artificial lighting.

(F): What did you choose between the functionality and aesthetics? How did you put it into practice?
(M): We didn't adopt only one of the functionality and aesthetics but integrated them into one whole. The decorations were chosen and made in accordance with the magnificent auras.

(F): What was the restriction that you should take care of when you design? What were the difficulties?
(M): The hardest thing was how to use the ordinary and common local materials to express another feeling which should differ from the local one. We were worried how if we made it like a local tea house. And the requirement of our client was changed in the later period of hard decoration from magnificent, casual, and

MATRIX DESIGN

cultural style to luxury style. The change was dramatic and we told a South Asia story by the soft decoration in later period to balance the change of the requirements of our client. It was finally done with original style that we preferred and also met the changed requirement of our client, which is to make it foreign luxury style.

(F): Did the local culture affect you on the design? If yes, how did you bring it into your design?
(M): Yes, and most of the materials that we used are of local feeling, which is plain, wild, dignified, and magnificent.

(F): Is there any design details that you like most in this project?
(M): There are 80 pendant lamps, on which there is sprays on the surfaces to reflect the outer river view. And this is the result of our designer's hard work in lamp factory for one week. The first version of the pendant lamp came out without meeting our designer's satisfactory and we broke it decisively.

(F): How did you get the design commission? What was the requirement of the commissioning party?
(M): We got the design commission in March 2011. At that time, our client required us to cooperate with a foreign design team to reconstruct an old and shabby Golf Club House into a New-Asian Style Sales Club House from the reconstruction of the building.

(F): Do you think if there was any conflict between the requirement of the commissioning party and the needs of the final consumer? If yes, how did you solve it and make it balance?
(M): There wasn't such conflict in this project. Because as the unique river view project, the Sales Club House also serves as Golf Club House. This was quite clear.

(F): In the process of the project, how did you communicate with the construction team to let them absolutely execute your design idea?
(M): This is a key question. We were lucky that the construction team of this project was quite cooperative and they had good execution ability. They constructed almost in accordance with the working drawings, and purchased the hard decoration material samples. However, I should point out that, in the design period, we designers should help the construction team to avoid some problems which are hard to solve, like the construction condition that the project can't meet, and the materials that exceed the cost of construction.

(F): Did the final result of the project after completion meet your expectation?
(M): Yes, it did. We can mark it 80 points for this project.

(F): 能具体谈谈您此次的获奖作品——"万科地产（重庆）天琴湾主会所"的设计理念吗？这个设计概念的起点是什么？
(M): 在整个地产行业充斥着米和金的时代里，我们渴望有一些新鲜的"空气"流入；如今外来文化到处侵占着我们的视野，于是我们期盼着中国的、东方的文化回归。"新亚洲"作为一股强劲的力量注入到我们的设计理念里，这也是整个项目定位的起点。空间凝重的大气场，对比其外部空间的开阔感，先抑后扬，注重材料表情的细腻表达，去表面化、去装饰化。

(F): 这个设计项目的秘密在哪里？是什么让它不同于一般的商业空间设计？
(M): 首先这是一个销售会所，去除贩卖的第一感受进而强调人与空间的关系，是我们一开始就与客户达成的共识，同时在注重身份认同感的同时强调人文休闲的气质。设计如果有"秘密"可言的话那就是：用心、用情去对待每一个项目，每一个甲方，每一个在空间里面的使用者。

(F): 这个设计项目您是否使用了什么特殊的材料或者独特的设计手法？
(M): 硬装几乎都是身边最常见而且代表地域特色的材料，将成本降至最低，软装则加入衬托风格需求的南洋韵味的材料，如贝壳、椰壳等。运用不同的排列组合来凸显其不一样的效果，夸张的尺度对比，材料表面的特殊处理加工，不管在自然光还是人造灯光的映衬之下，都能呈现出独特的质地。

(F): 您对功能性和美观性这两方面如何选择？如何把这种选择付诸实践？
(M): 功能与大空间韵律统一，装饰手法与大气场的表情统一。

(F): 在设计的过程中有什么限制条件是您必须处理的？遇到的困难是什么？
(M): 最艰难的问题在于，如何使用最普通而常见的本地材料来表达出不同于本地的风情感受。当时很担心做出来的东西会跟当地的茶楼一样，而且客户在硬装后期的要求发生改变；从一开始的凝重大气、休闲人文，改变为直接明了的显贵。这个变化是很有戏剧性的，我们通过后期的软装讲述了一个南洋的故事，从而平衡了一系列的要求，既满足了原有格调的统一延续，又满足了营销对外来文化的推崇。

(F)：当地文化对您此次的设计有影响吗？您如何将它们带入您的设计中？
(M)：材料表情几乎都是延续本土的基调：质朴粗犷、浑厚大气。

(F)：这个设计项目有没有什么您最钟爱的设计细节？
(M)：比如软装配饰中80多盏吊灯，上面浮雕着呼应外部江景的浪花纹样，这是我们自己的设计师在灯具厂花费一周的心血完成的。第一版被我们毫不留情地砸掉了。

(F)：这个设计项目是如何发生的？委托人的需求是什么？
(M)：2011年3月，当时客人的要求是让我们配合境外的设计团队从建筑改造开始，将一个原有很破旧的高尔夫会所改造成新亚洲风格的销售会所。

(F)：您认为委托人的需求和最终消费者的需求是否有冲突？如果有，您如何平衡、解决这种冲突？
(M)：在这个项目中基本不存在这样的问题，因为这个项目是仅有的江景项目，销售会所同时服务于高尔夫会所的经营需要，其本身的身份对等感已经十分清晰了。

(F)：在项目进行的过程中，您如何与施工团队沟通，使他们完全执行您的设计想法？
(M)：这是个很关键的问题点，很有幸这次的施工团队的配合度比较高，执行力很强，基本都是按图施工，按材料样板采购的。但同时我想讲的是，我们在设计前期的时候，设计师应该帮助施工队在有限的施工期中屏蔽掉一些很难执行的问题点，比如说不符合本地施工的节点，不符合造价的材料。

(F)：这个项目竣工后的成果符合您的预期吗？
(M)：如果要打分的话已经可以打80分啦！

Process 心路历程

A Feb. 14th, 2011
Site Visit
20110214 现场勘察

B Apr. 8th, 2011
Reconstruction of the Building
20110408 建筑改造中

C May. 29th, 2011
On-site Communication Before Construction
20110529 施工交底

E Jul. 3rd, 2011
Pictures in Process
20110703 过程中照片

D Aug. 18th, 2011
Sample Production of Furniture
20110818 家私打版

F Sep. 28th, 2011
Sidelights of Soft Decoration Placement
20110928 软装摆场

Process 心路历程

Feb. 14th, 2011
Site Visit

20110214 现场勘察

这栋楼的原貌

早春的天气还是很寒冷

拆除的楼梯

Apr. 8th, 2011
Reconstruction of the Building

20110408 建筑改造中

天气转暖已经可以穿皮衣啦

大堂一角

地下层水吧的位置

May. 29th, 2011
On-site Communication Before Construction

20110529 施工交底

外墙改造已经开始

这个时候正值夏日

拆空后的大堂

Jul. 3rd, 2011
Pictures in Process

20110703 过程中照片

外观改造已近完成

盛夏时节工人师傅都很辛苦

地下层的水吧台

Aug. 18th, 2011
Sample Production of Furniture

20110818 家私打版

与甲方研究工艺

我们的设计师耗时一周雕刻的白胚成型

检查纹理

MATRIX DESIGN

Process 心路历程

的过程中发现了些问题

现场对图纸

与SCDA的设计师共同商讨

原来的大堂挑空部分

的餐厅空间

改造后一层水吧的位置

洽谈区的末端位置

三角形的洞口就是新的楼梯的位置

端倪的旋转梯

大堂局部的天花还未拆改完成

二层加建的天花

新建楼梯的雏形

层与上部空间的挑空部分

楼梯的扶手还没安装

洽谈区正在铺装地面

外墙正在饰面

的底座

两个模板

整木的效果

实实在在的全实木家私白胚

Process 心路历程

The project took half a year, from accepting the commission in early Spring to completion of the project in late Autumn. We are thankful to enjoy a harvest of this wonderful project in this harvest time. We are also thankful to our happily working team which is full of passion. We'd like to share some stories with you, which are actually the happy sidelights of the hard work in placement of the soft decoration. Meanwhile, we'd like to send our thankfulness to our clients and the colleagues that support us.

此项目历时半年，从接受委托到交付完工经历了初春到深秋。能在收获的季节收获如此作品，离不开一个有激情并快乐工作的团队。我们将软装摆场过程中苦中作乐的花絮，编成了几个图片小故事与大家分享。同时也感谢支持我们工作的甲方及同仁们。

Process 心路历程

298/299

Process 心路历程

MATRIX DESIGN

Process 心路历程

完工后现场实景拍摄

没错!就是这六个屌丝干的好事!

图书在版编目（CIP）数据

矩阵纵横设计作品精选 / 矩阵纵横设计团队主编
. -- 南京：江苏人民出版社，2013.1
ISBN 978-7-214-08958-8

Ⅰ．①矩… Ⅱ．①矩… Ⅲ．①室内装饰设计—作品集
—中国—现代 Ⅳ．①TU238

中国版本图书馆CIP数据核字(2012)第279246号

矩阵纵横设计作品精选

矩阵纵横设计团队　主编

责任编辑	刘　焱
特约编辑	陈丽新
版式设计	魏中友
封面设计	魏中友
责任监印	安子宁
出版发行	凤凰出版传媒股份有限公司
	江苏人民出版社
	天津凤凰空间文化传媒有限公司
销售电话	022-87893668
网　　址	http://www.ifengspace.cn
经　　销	全国新华书店
印　　刷	利丰雅高印刷（深圳）有限公司
开　　本	965 mm×1270 mm　1/16
印　　张	19
字　　数	152千字
版　　次	2013年1月第1版
印　　次	2013年1月第1次印刷
书　　号	978-7-214-08958-8
定　　价	258.00(USD 53.00)

（本书若有印装质量问题，请向销售部调换）